ON CARING

ON CARING

MILTON MAYEROFF

PERENNIAL LIBRARY
Harper & Row, Publishers
New York, Evanston, San Francisco, London

This PERENNIAL LIBRARY paperback edition is a reprint of Volume XLIII of the WORLD PERSPECTIVES series, which is planned and edited by RUTH NANDA ANSHEN. Dr. Anshen's Epilogue to this reprint appears on page 89.

A hardcover edition of this work is published by Harper & Row, Publishers.

ON CARING

First PERENNIAL LIBRARY edition published 1972.

STANDARD BOOK NUMBER: 06-080242-1

6-74

IN MEMORY OF MY FATHER AND MOTHER

JACOB AND BERTHA MAYEROFF

Contents

Acknowledgments

I would like to thank four friends, Justus Buchler, Edwin A. Burtt, Albert Hofstadter, and Moreland Perkins, for encouragement and helpful criticism.

My thanks also to the Research Foundation of the State University of New York for three Faculty Research Fellowships.

I am especially indebted, in very different ways, to the writings of John Dewey, Erich Fromm, Gabriel Marcel, and Carl R. Rogers.

My thanks are due to the Editor of *The International Philosophical Quarterly* for permission to use some of the material which appeared in my article, "On Caring," in September, 1965.

. . . the degree to which I can create relationships which facilitate the growth of others as separate persons is a measure of the growth I have achieved in myself. . . .

—Carl R. Rogers

A man cannot be a true father without being a creator of something, something that he wants to flourish before him rather than to have molded in his image. A true father must be an artist, otherwise he becomes a violator of the creative spirit of man, by trying to mold a child to be exactly the way he the man is. . . .

—Gregory Zilboorg

The thematic idea, *this child of truly creative musicianship, is a live birth and has individual qualities; and in his treatment, shaping, and development of it, its progenitor should have regard for the child's nature, and not proceed in an arbitrary fashion. The true composer does not behave like a tyrant towards his thematic substance; but he watches like a provident father for signs suggestive of individual development, and lets his creative phantasy be fertilized and directed by these.*

—Bruno Walter

A book takes on its own life in the writing. It has its laws, it becomes a creature to you after a while. One feels a bit like a master who's got a fine animal. Very often I'll feel a certain shame for what I've done with a novel. I won't say it's the novel that's bad; I'll say it's I who was bad. Almost as if the novel did not really belong to me, as if it was something raised by me like a child. I know what's potentially beautiful in my novel, you see. Very often after I've done the novel I realize that that beauty which I recognize in it is not going to be recognized by the reader. I didn't succeed in bringing it out. It's very odd—it's as though I had let the novel down, owed it a duty which I didn't fulfill.

—NORMAN MAILER

I don't want to sound metaphysical, but there arrives a period during the painting when the painting itself makes certain demands, you see, and if you're not hypersensitive to it, you know, you're going to lose a good quality of painting. It definitely becomes a living thing.

—BEN SHAHN

We ourselves want to be needed. We do not only have needs, we are also strongly motivated by neededness. . . . *We are restless when we are not needed, because we feel "unfinished," "incomplete," and we can only get completed in and through these relationships. We are motivated to search not only for what we lack and need but also for that for which we are needed, what is wanted from us.*

—ANDRAS ANGYAL

For only to the extent that man has fulfilled the concrete meaning of his personal existence will he also have fulfilled himself. . . . The meaning which a being has to fulfill is something beyond himself, it is never just himself.

—VIKTOR E. FRANKL

ON CARING

1. *Introduction*

To care for another person, in the most significant sense, is to help him grow and actualize himself. Consider, for example, a father caring for his child. He respects the child as existing in his own right and as striving to grow. He feels needed by the child and helps him grow by responding to his need to grow. Caring is the antithesis of simply using the other person to satisfy one's own needs. The meaning of caring I want to suggest is not to be confused with such meanings as wishing well, liking, comforting and maintaining, or simply having an interest in what happens to another. Also, it is not an isolated feeling or a momentary relationship, nor is it simply a matter of wanting to care for some person. Caring, as helping another grow and actualize himself, is a process, a way of relating to someone that involves development, in the same way that friendship can only emerge in time through mutual trust and a deepening and qualitative transformation of the relationship. Whatever the important differences are among a parent caring for his child, a teacher caring for his pupil, a psychotherapist caring for his patient, or a husband caring for his wife, I would like to show that they all exhibit a common pattern. But besides caring for people, in this sense, we may care for many other things as well. We may care, for instance, for our "brain child" (a philosophical or an artistic idea), an ideal, or a community. And here too, whatever the important differences are between caring for a person and caring for an idea, I would like to show that there is a common pattern of helping the other grow. It is this general pattern of caring that I will describe and explore.

In the context of a man's life, caring has a way of ordering his other values and activities around it. When this ordering is comprehensive, because of the inclusiveness of his carings, there is a basic stability in his life; he is "in place" in the world, instead of being out of place, or merely drifting or endlessly seeking his place. Through caring for certain others, by serving them through caring, a man lives the meaning of his own life. In the sense in which a man can ever be said to be at home in the world, he is at home not through dominating, or explaining, or appreciating, but through caring and being cared for.

This small book deals with these two related themes: a generalized description of caring and an account of how caring can give comprehensive meaning and order to one's life. The two concepts, "caring" and being "in place," provide a fruitful way of thinking about the human condition; and, what is more important, they may help us understand our own lives better. Much that is important about man cannot be illuminated by these concepts, but I believe they help us understand something of what is most important.

I

Caring as Helping the Other Grow

2. The Basic Pattern

In caring as helping the other grow, I experience what I care for (a person, an ideal, an idea) as an extension of myself and at the same time as something separate from me that I respect in its own right. This feeling of the other as part of me is different from the kind of union with the other found in such parasitic relations as morbid dependence on another person or dogmatically clinging to a belief, for in both these cases I am unable to experience the other as independent in its own right and I am unable to respond to it truly.* When I dogmatically cling to a belief,

* To bring out the general applicability of caring, I will speak of caring for *it* except where it is apparent that I am referring only to caring for a person.

I am so attached to it that I am unable to experience it as separate from me, and I cannot really examine the belief and find out what it means, let alone determine whether it is true or false.

The union with the other in caring differs in another way from that found in a parasitic relation. Instead of trying to dominate and possess the other, I want it to grow in its own right, or, as we sometimes say, "to be itself," and I feel the other's growth as bound up with my own sense of well-being. The worth I experience in the other is something over and above any value it may have for me because of its ability to satisfy my own needs. For a caring parent, the child is felt to have a worth of his own apart from his power to satisfy the parent's needs; for the caring musician, the music is felt to have a worth of its own apart from what it is able to do for him. In other words, I experience what I care for as having worth in its own right.

In caring I experience the other as having potentialities and the need to grow; I experience an idea, for instance, as seminal, vital, or promising. In addition, I experience the other as needing me in order to grow; consider how we sometimes feel needed by another person or by a cause or an ideal. This does not simply mean that I know, in some strictly intellectual sense, that the other has needs that must be satisfied and that I can satisfy those needs. And I do not experience being needed by the other as a relationship that gives me power over it and provides me with something to dominate, but rather as a kind of trust. It is as if I had been entrusted with the care of the other in a way that is the antithesis of possessing and manipulating it as I please.

To say the artist experiences his "spiritual child" as having a life of its own, and as striving to grow and needing him in order to grow, does not mean he believes it is conscious and has feelings. This simply describes how he experiences the work of art. It is why we speak of respecting the integrity of the artwork. We engage in a similar personification when we see a house as having a "personality," or experience a house as a home, but it does not distort our perception by making for unreasonable expectations. Yet this is not to deny that the work of art could not grow without the artist's help, or that outside some human context it would make no sense to speak of an artwork as alive and striving to grow.

In helping the other grow I do not impose my own direction; rather, I allow the direction of the other's growth to guide what I do, to help determine how I am to respond and what is relevant to such response. I appreciate the other as independent in its own right with needs that are to be respected: as we sometimes say in the context of inquiry, "We follow the lead of the subject matter."

Direction that comes from the growth of the other should not be confused with being "other directed," where this refers to the kind of conformity in which I lose touch with both myself and the other. Rather, by following the growth of the other, I am more responsive to myself, just as the musician is more in touch with himself when he is absorbed in the needs of the music. Any direction that I may give the other is governed by my respect for its integrity and is intended to further its growth, and I show that respect by the interest I take in determining whether my actions do in fact further growth and by being guided by what I find.

Devotion is essential to caring, just as it is an integral part of friendship. I commit myself to the other and to a largely unforeseeable future. Devotion is not an element that may or may not be present, as if I might be said to care and *also* to be devoted. When devotion breaks down, caring breaks down. Again, devotion does not simply measure the extent of my caring, but it is through devotion that caring for *this* other acquires substance and its own particular character; caring develops in the process of overcoming obstacles and difficulties. My devotion is grounded in the worth I experience in the other. Such devotion expresses my entire person rather than simply the intellectual or emotional part of me. Viewed at a particular time, devotion is shown by my being "there" for the other in a way that is the converse of holding back and ambivalence. Viewed over an extended period, it is shown by my consistency, which expresses itself in persistence under unfavorable conditions, and in my willingness to overcome difficulties. This consistency is liberating and is an expression of my will, unlike the rigid consistency of compulsive behavior which enslaves me and makes me feel driven and imposed on by something outside myself.

Obligations that derive from devotion are a constituent element in caring, and I do not experience them as forced on me or as necessary evils; there is a convergence between what I feel I am supposed to do and what I want to do. The father who goes for the doctor in the middle of the night for his sick child does not experience this as a burden; he is simply caring for the child. Similarly, in working out a philosophical concept the need to reflect on it again and again from similar and dissimilar points of view is not a burden forced on me; I am simply caring for the idea.

This, then, is the basic pattern of caring, understood as helping the other grow: I experience the other as an extension of myself and also as independent and with the need to grow; I experience the other's development as bound up with my own sense of well-being; and I feel needed by it for that growing. I respond affirmatively and with devotion to the other's need, guided by the direction of its growth. This pattern will be subsequently clarified and developed by describing its major ingredients and certain illuminating aspects. To bring out the generality of the pattern, I often speak of caring for the *other,* but in any actual instance of caring it is always someone or something specific that is cared for: the writer cares for *this* idea, the parent cares for *this* child, the citizen cares for *this* community.

3. The Growth of a Person and the Growth of an Idea

Although it is not my purpose to examine specifically the nature of growth and actualization, since caring is understood as helping the other grow, some brief remarks about the growth of a person and the growth of a philosophical idea like "caring" will be helpful for our discussion. The remarks about the growth of a person which follow are rather general; on a more concrete level there would be differences depending on whether we referred specifically to the growth of a child to maturity, the growth of an immature adult to maturity, or the growth of a mature adult.

To help another person grow is at least to help him to care for something or someone apart from himself, and it involves encouraging and assisting him to find and create areas of his own in which he is able to care. Also, it is to help that other person to come to care for himself, and by becoming responsive to his own need to care to become responsible for his own life. Growing includes learning to the degree that one is able, where learning is to be thought of primarily as the re-creation of one's own person through the integration of new experiences and ideas, rather than as the mere addition of information and technique. I grow by becoming more self-determining, by choosing my own values and ideals grounded in my own experience, instead of either simply conforming to the prevailing values or compulsively rejecting them. I am better able to make my own decisions and more willing to be responsible for them, and I can discipline and limit myself in order to seek out and achieve what is important to me. A man grows by becoming more honest with himself and more aware of the social and natural order of which he is a part; in coming to see himself with a minimum of illusion, he also comes to appreciate better the objective structure of means and ends.

I help a philosophical idea like "caring" grow by discovering and exploring its essential traits. Its scope of application is widened and clarified, and various activities, seemingly disparate (for example, child rearing, teaching, painting, psychotherapy), turn out to be related as they are shown to provide the opportunity for caring. Along with growth in generality, there is also growth in specificity: the ideas help us to understand better particular activities, for example, to realize new and interesting things about teaching and art making. The concept of "caring" is developed by disclosing its relationships to other significant concepts like "trust," "honesty," and "humility," and it also grows through coming to terms with seeming exceptions. From a loose stringing together of ideas, a tight fabric emerges; ideas intertwine and tend to reinforce each other, making for a mutual deepening of meaning and a gain in precision. With the growth of an idea comes a deeper understanding of what its basic assumptions are, what it can and cannot do, and a clear sense of what is relevant and irrelevant for its further development. I help a concept like "caring" grow not only by discovering and describing its essential ingredients, but also by considering it in a wider context and seeing how it functions and could function within a man's life viewed as a whole.

II

Major Ingredients of Caring

4. Knowing

We sometimes speak as if caring did not require knowledge, as if caring for someone, for example, were simply a matter of good intentions or warm regard. But in order to care I must understand the other's needs and I must be able to respond properly to them, and clearly good intentions do not guarantee this. To care for someone, I must *know* many things. I must know, for example, who the other is, what his powers and limitations are, what his needs are, and what is conducive to his growth; I must know how to respond to his needs, and what my own powers and limitations are. Such knowledge is both general and specific. The composer, for example, does not simply know about music in general, nor does he simply know about a specific musical idea; in order to care he must have both sorts of knowledge. General and specific knowledge go hand in hand: the composer, for instance, uses his general knowledge of composing to care for a particular piece of music, and by caring for that piece of music he learns more about composing in general.

What we know in caring, we know in different ways. We know some things *explicitly* and some things *implicitly.* To know something explicitly is to be able to tell what we know, to be able to put it into words; by contrast, to know something implicitly is to be unable to articulate it. We know more about a good friend than we can verbalize. Second, there is the difference between knowing *that* something is so and knowing *how* to do something. A man may know much about the theory of teaching without being able to teach. Third, there is the difference between *directly* and *indirectly* knowing something. By knowing something directly, I mean encountering it, apprehending it as existing in its own right; I do *not* mean simply experiencing it. In caring, I know the other directly; the union I experience with the other goes with my awareness of its separateness and individuality. The caring teacher, for example, directly knows his student as an individual: he experiences him as someone in his own right, and not as a stereotype or as a means for his own self-aggrandizement. By contrast, indirect knowledge refers to knowing about something, to having information about it. I may know something indirectly without actually experiencing it, and I may experience it without knowing it directly.

Caring, then, includes explicit and implicit knowledge, knowing that and knowing how, and direct and indirect knowledge, all related in various ways to helping the other grow. One important reason, perhaps, for our failure to realize how much knowing there is in caring is our habit sometimes of restricting knowledge arbitrarily to what can be verbalized. We do not consider implicit knowledge, knowing how, and direct knowledge as ways of knowing. Restricting the meaning of knowledge in this way is as arbitrary as assuming that only words can be communicated and restricting the meaning of communication to what can be put into words.

5. Alternating Rhythms

As a teacher I try to explain some idea to a student, look to see whether I have succeeded, and if I have not, try again in some other way. Or as a writer I try to put a thought into words, read it over to see whether I have succeeded, and if I have not, try again in some other way. In both cases I act with certain expectations, undergo or suffer the results of my actions, and then link up these two phases and see whether what I expected was in fact achieved. I cannot care by sheer habit; I must be able to learn from my past. I see what my actions amount to, whether I have helped or not, and, in the light of the results, maintain or modify my behavior so that I can better help the other. But "doing" is to be understood broadly and not only in the active sense, as though I were always acting on the other. It may involve doing "nothing." In caring for a person, for instance, there are times when I do not inject myself into the situation, I do not take a stand one way or the other, I do "nothing." And when I undergo this "inactivity," I see what resulted from it and may change my behavior accordingly.

Consider a different sort of rhythm that is also important in caring, the rhythm of moving back and forth between a narrower and a wider framework. There are times in caring for a child when I examine an act as a relatively isolated episode, without relating it carefully to what went before and what will follow, and other times when I look at the particular act in its wider connections within a larger framework and can discern trends, long-term effects, and tendencies. It is one thing, for example, to examine an act of insecurity as a relatively isolated event; it is another matter to consider it as the expression of a general pattern of insecurity. Or, in working out an idea, there are times when I attend to a detail in relative isolation, and times when I view the detail in connection with other ideas or with the projected essay or book as a whole. To see the way a chapter fits into a projected book may change my idea of what the book is to be, or it may change my idea of what the chapter ought to be. And clearly, the chapter we view in relative isolation at one time may, at another time, itself provide a wider context for examining some page.

6. Patience

Patience is an important ingredient in caring: I enable the other to grow in its own time and in its own way. (The growth of a significant idea can no more be forced than the growth of a flower or a child.) By being patient I give time and thereby enable the other to find itself in its own time. The impatient man, on the other hand, not only does not give time, but he often takes time away from the other. If we know that someone is impatient with us, or if we are impatient with ourselves, even the time that we might have had is often reduced.

Patience is not waiting passively for something to happen, but is a kind of participation with the other in which we give fully of ourselves. And it is misleading to understand patience simply in terms of time, for we give the other space as well. By patiently listening to the distraught man, by being present for him, we give him space to think and feel. Perhaps, instead of speaking of space and time, it would be truer to say that the patient man gives the other room to live; he enlarges the other's living room, whereas the impatient man narrows it.

Patience includes tolerance of a certain amount of confusion and floundering. But this tolerance is not adherence to a rule which says I ought to be tolerant, nor is it a kind of indifference to the other. Rather, tolerance expresses my respect for the growth of the other, and my appreciation of the "wastefulness" and free play that characterize growth.

The man who cares is patient because he believes in the growth of the other. But, besides being patient with the other, I must also be patient with myself. I must give myself a chance to learn, to see and to discover both the other and myself; I must give myself a chance to care.

7. *Honesty*

Honesty is present in caring as something positive, and not as a matter of *not* doing something, not telling lies or not deliberately deceiving others.

This meaning of honesty may be suggested by the phrase, "to be honest with oneself," where this includes actively confronting and being open to oneself. In caring I am honest in trying to see truly. To care for the other, I must see the other as it is and not as I would like it to be or feel it must be. If I am to help the other to grow, I must respond to *its* changing needs. If I have to see the other in a certain way, if I can see only what I would like to see, I will not be able to see the other as it really is. Idolatry, for instance, militates against caring, for it makes it impossible really to respond to *this* other. Even when the facts are unpleasant, I respect them, for it is only by taking them seriously that I can be in touch with the other and care for it. But besides seeing the other as it is, I must also see myself as I am: I must see what I am doing and whether what I am doing helps or hinders the growth of the other. The writer must have the courage to look and see whether he is more interested in proving himself correct than he is in examining and developing the idea, or whether he is more interested in having something published than he is in developing the idea.

I can be honest and still be mistaken. But I try, and where I am wrong, I am open to correction and try to learn from my mistakes. My desire to help the other militates against my having a stake in maintaining distortion. I am honest in caring not because of expediency ("Honesty is the best policy"), as if honesty were only a means to caring, but because honesty is integral to caring.

Honesty is also present in caring in a different

way. I must be genuine in caring for the other; I must "ring true." There must not be a significant gap between how I act and what I really feel, between what I say and what I feel. To be "present for" the other, so that the other can be present for me, I must be open to the other. Pretending to be what I am not interferes with being able to relate to the other as an individual in its own right; I cannot be fully present for the other if I am more concerned about how I appear to other people than I am with seeing and responding to its needs. The parent who must constantly prove how much he cares gets in the way of caring for his child.

8. Trust

Caring involves trusting the other to grow in its own time and in its own way. It appreciates the independent existence of the other, that the other is *other*. In caring for another person I trust him to make mistakes and to learn from them. We trust the child *now* to make those decisions for himself that are reasonably commensurate with his experience and his abilities. The realization that "he trusts me" has its own way of activating the person cared for to justify such trust and to trust himself to grow. In working out ideas we show trust in following their lead and allowing organization to emerge from them. We also show trust in letting them "come home" to us so that we may understand what it is we are doing, and in ultimately exposing them to examination and criticism by others. Trusting the other is to let go; it includes an element of risk and a leap into the unknown, both of which take courage.

We show lack of trust by trying to dominate and force the other into a mold, or by requiring guarantees as to the outcome, or even by "caring" too much. Insofar as schooling or religious instruction is primarily indoctrination without allowing the opportunity to question and reflect on what is taught, it is rooted in lack of trust in the other. The man who fears and avoids the unknown, who must always be sure how it will all turn out, cannot allow the other to grow in its own way. He becomes unresponsive to the needs of the other.

The father who "cares" too much and "overprotects" his child does not trust the child, and whatever he may think he is doing, he is responding more to his own needs than to the needs of the child to grow. He does not see the child as having the need to be independent and to be responsible for himself. Morbid dependency by its very nature is incompatible with trust, for in such a situation any sign of independence on the part of the other is experienced as a threat.

Trust in the other to grow is not indiscriminate; it is grounded in actively promoting and safeguarding those conditions which warrant such trust. The caring teacher who trusts his students to find their own way in pursuing their own projects grounds such trust by providing the students with assistance, encouragement, and exposure to relevant and stimulating experiences. But only the man who trusts himself to grow, who is not trying to force himself to be something he thinks he is supposed to be, will be able to trust another person to grow.

Besides trusting the other, I must also trust my own capacity to care. I must have confidence in my judgments and in my ability to learn from mistakes; I must, as we say, trust my instincts. The philosophical writer must trust his feelings for importance and relevance, for determining when ideas ring true and when to leave them out. The teacher must trust his ability to provide a climate friendly to learning, and to learn from student reactions what works and what does not. And the parent must trust his judgment to know when firmness is required, and to know when seemingly isolated instances really point to more lasting habit patterns. Continuing preoccupation with whether my actions are correct indicates lack of trust in myself and, in focusing attention on myself, makes for further indifference to the needs of the other.

9. Humility

Humility is present in caring in several ways. First, since caring is responsive to the growth of *this* other, caring involves continuous learning about the other: there is always something more to learn. The man who cares is genuinely humble in being ready and willing to learn more about the other and himself, and what caring involves. This includes learning from the one cared for as well: the teacher learns from the student; the parent learns from the child; and the artist learns from the work of art. No source is felt to be beneath me in principle; I am not humiliated to learn from any source, including my own mistakes. An attitude of not having anything further to learn is incompatible with caring. The father who already "knows" his child completely, the "patriot" who feels he can learn nothing about his country from other people, do not care. Also, there is a sense in which the man who cares basically begins anew regardless of how extensive his previous experience has been, for the problem is always one of appropriateness to *this* novel situation, and this situation is, generally speaking, not simply a repetition of the past requiring only a mechanical application of principles.

Humility is also present in realizing that my particular caring is not in any way privileged. What is ultimately significant is not whether my caring is more important than your caring, but that a man be able to care and have something to care for. The preoccupation with whose caring is worth more takes me away from caring. I become more concerned about myself and the fact that it is *I* who do the caring than I am about the growth of the other.

In addition, caring itself expresses a broader meaning of humility as the overcoming of an attitude that sees others as existing simply to satisfy my own needs, and treats others as if they were merely obstacles to overcome or clay for me to mold as I please. It includes overcoming the arrogance that exaggerates my own powers at the expense of the powers of others, and blinds me to the extent of my dependence, in anything I accomplish, on the cooperation of various conditions over which I have little or no control. Humility also means overcoming pretentiousness: I am able to present myself as I am without self-display and concealment, without posing and indirection. And since I do not pretend to be what I am not, I am not humiliated by having others see me truly: in being open there is nothing for others to see through. Caring expresses this broader meaning of humility because it recognizes that others have an integrity of their own.

Through caring I come to a truer appreciation of my limitations as well as my powers; my limitations are neither resented nor glorified, and I can take pride in the successful use of my powers. Consider, for example, the mother's pride in realizing how she has helped her child grow in independence and responsibility, or the philosopher's pride in the thoroughness with which he has worked out a significant idea. Such pride is very different from vanity, and does not include vindictive triumph over another person—there is nothing arrogant about it. Instead of dividing me from others, it opens me more to the world and puts me more in touch with myself and others. Pride in a job well done is not pretentious, it does not distort; rather it goes with an honest awareness of what I have done and the extent of my dependence on the cooperation of others and on various conditions. There is nothing incompatible between pride, in this sense, and humility.

10. Hope

There is hope that the other will grow through my caring which is more general than hope as a specific expectation; it is akin, in some ways, to the hope that accompanies the coming of spring. It is not to be confused with wishful thinking and unfounded expectations. Such hope is not an expression of the insufficiency of the present in comparison with the sufficiency of a hoped-for future; it is rather an expression of the plenitude of the present, a present alive with a sense of the possible. For example, in caring for a child I am stirred by the possibilities to be realized, and this is bound up with my hope for the growth of the child through my caring. By contrast, where there is no possibility of new growth, there is despair.

Hope's reference to the future in caring enlarges the significance of the present; it does not subordinate the present to something lying beyond it and turn it into a mere means. The father who is unable to trust his child as someone in his own right may have great "hopes" for the child, but they have little to do with the awareness of *this* child now. Such hopes actually impoverish the present by making it largely a postponement for a "more real" future when the child will really "amount" to something.

Hope, as an expression of a present alive with possibilities, rallies energies and activates our powers; it is not a passive waiting for something to happen from outside. But it is not simply hope for the other, it is hope for the realization of the other *through* my caring; and therefore an important aspect of hope is courage. Such courage is found in standing by the other in trying circumstances, and in taking risks that go beyond safety and security. If I did not believe that I would stand up for the other in difficult circumstances, my hope for the growth of the other through my caring would be necessarily undermined. But not only does courage make hope possible, it is equally true that hope makes for courage; for hope implies that there is or could be something worthy of commitment. Lack of hope, on the other hand, eats away any sense of worthiness, and therefore anything for which I would want to take a stand. Put differently, despair militates against courage, it drains it of vitality.

11. Courage

Courage is also present in going into the unknown. By following the lead of the subject matter or the direction of the growing child, I have no guarantee where it will all end or in what unfamiliar situations I will find myself. The security of familiar landmarks is gone and I cannot anticipate fully who or what the other will become or who I will become. This is the courage of the artist who leaves the fashions of the day to go his own way, and in so doing comes to find himself and be himself. Such courage is not blind: it is informed by insight from past experiences, and it is open and sensitive to the present. Trust in the other to grow and in my own ability to care gives me courage to go into the unknown, but it is also true that without the courage to go into the unknown such trust would be impossible. And clearly, the greater the sense of going into the unknown, the more courage is called for in caring.

III

Some Illuminating Aspects of Caring

12. Self-Actualization Through Caring

In caring, the other is primary; the growth of the other is the center of my attention. The teacher's interest is focused on the student rather than on himself. To make himself the center of his attention would get in the way of his caring for the student. Only by focusing on the other am I able to be responsive to its need to grow.

There is a *selflessness* in caring that is very different from the loss of self in panic or through certain kinds of conformity. It is like the selflessness that goes with being absorbed in something I find genuinely interesting, that goes with being "more myself." Such selflessness includes heightened awareness, greater responsiveness to both the other and myself, and the fuller use of my distinctive powers.

In caring for the other, in helping it grow, I actualize myself. The writer grows in caring for his ideas; the teacher grows in caring for his students; the parent grows in caring for his child. Or, put differently, by using powers like trust, understanding, courage, responsibility, devotion, and honesty I grow also; and I am able to bring such powers into play because my interest is focused on the other.

Besides the other's need for me if it is to grow, I need the other to care for if I am to be myself. The teacher needs his student, just as the student needs the teacher. The philosopher needs his seminal idea, just as the idea "needs" the philosopher. But to say I need the other if I am to be myself does not mean I experience the other as a means, as existing simply to satisfy my own needs. *I do not try to help the other grow in order to actualize myself, but by helping the other grow I do actualize myself.* My dependence on the other is bound up with respecting and furthering its integrity, which is very different from a parasitic relation in which I want to possess the other and am unable to experience it as existing in its own right. And this does not deny the possibility of conflict between caring for some other and caring for myself.

13. The Primacy of the Process

The process rather than the product is primary in caring, for it is only in the present that I can attend to the other. The problem is always how to respond to this person or this idea, here and now; we must always work, so to speak, with what we have and from where we are. Control is possible only in the present. General impatience with the process and the wish to eliminate it entirely show ignorance of what growth is all about. In caring the present is not cut off from vital connections with the past and future, for it is informed by meanings and insights from the past and enriched by anticipations of the future, such as the promise of new growth. But at the same time that past and future make us more sensitive to opportunities for growth in the present, the interests and needs of the present help determine the general character of this past and future. They help determine which past meanings and insights are relevant to the present, and which possibilities are real and significant for this present.

To speak of the primacy of the process is not to deny the important role which anticipated goals and general aims play; they contribute to direction and meaning in the present. Consider, for instance, how the artist's vision of the completed work or the psychotherapist's views about cure function in the present. But if the present (the process) is not taken seriously for its own sake, and is basically subordinated to the future (the product) by being treated either as a necessary evil or as a mere means to something lying beyond, then caring becomes impossible. The father impatient for his child to grow up and become "something" that he is not now does not really take the child seriously and makes caring impossible.

Also, real interest in the outcome expresses itself in the concern we take *now*, for the product is an outgrowth of the process, or, put differently, the process is the product in the making. In other words, the test of the genuineness of my concern for the future (a present that is yet to come) is the care I take with what is present now. The primacy of the process may be illustrated in another way. We have seen that caring involves self-actualization for the one who cares. Thus a writer grows in the course of working out and caring for his ideas and not, presumably, when his book is finished. After his book is finished and the accompanying satisfactions are over, the really important question is: "What am I to care for now?"

14. The Ability to Care and the Ability to Be Cared For

Caring sometimes calls for unusual aptitudes and special training; besides being able to care in general, I must be able to care for this specific other. Caring for a mentally ill person requires uncommon sensitivity in interpersonal relations as well as specialized training; caring in performing the late Beethoven piano sonatas requires deep musical understanding as well as an accomplished technique.

If I am to care for the other, I must be able to cope with it; I must be "up to" caring for it. It is not enough merely to want to care for the other and desire its growth; I must be able to help it grow. And just as I must be capable of caring for *this* other, this other must be capable of being cared for. The psychotherapist cannot help his patient grow if the patient does not, at least on some level, really want to grow. We sometimes say about a person, "He doesn't give anyone a chance to help him." If a person has suffered extensive brain damage and is unable to grow in any meaningful sense, I may comfort him and be interested in his welfare, but I cannot care for him in the sense of helping him to grow. Similarly, some ideas lack the capacity to grow, they lack depth; we speak of them as "dead" or "wooden." They cannot be cared for.

15. The Constancy of the Other

Caring assumes continuity, and is impossible if the other is continually being replaced. The other must remain constant, for caring is a developmental process. In situations of great social mobility man becomes rootless, and loyalty to one's community becomes increasingly difficult. We do not stay in one place long enough for any sense of loyalty to develop; powers like devotion and trust do not get a chance to come into play. Friendship is a relationship in depth that takes its own time to develop and is not possible if the other person is continually being replaced. But the point is not simply that the other must not in fact be replaced, but that in caring, in committing ourselves, we feel the other to be constant, to be there for us to help. The other not only must be constant but must be experienced as constant.

16. Guilt in Caring

In caring I commit myself to the other; I hold myself out as someone who can be depended on. If there is an acute break within this relation because of my indifference or neglect, I feel guilty, as if the other were to say, "Where were you when I needed you, why did you let me down?" This guilt results from my sense of having betrayed the other, and my conscience calls me back to it. The more important this particular other is to me, the more pronounced is my guilt.

Like pain, guilt tells me that something is wrong; if it is felt deeply, understood, and accepted, it provides me with the opportunity to return to my responsibility for the other. That return does not necessarily reinstate the relationship as it existed prior to the break; rather, it often makes for a deeper seriousness and awareness of my trust. It is like almost losing something through indifference, and by this near-loss realizing more deeply how precious it actually is to me. I do not resume caring simply to overcome guilt, but I overcome guilt by renewed caring.

Also, since I identify with the growth of the other, and experience it as in some sense an extension of myself, my neglect of it produces at the same time a break in my own responsiveness to myself. Just as the honorable man betrays himself in breaking his word to another, guilt in caring is not simply an expression of my betrayal of the other; it is also an expression of self-betrayal. Conscience calls me back both to the other *and* to myself. Through overcoming the break with the other, I overcome the break within myself.

17. Reciprocation

Caring may or may not be reciprocated. Things cannot respond to me as I respond to them; their "personality" has largely been given to them by me. A work of art, for example, obviously cannot care for me in the sense in which I have been speaking. Although I may feel myself in good hands while listening to a Bach cantata, to think of it as caring for me would be to stretch inordinately the concept of caring. The relationship between a caring parent and his very young child is another example of caring that cannot be reciprocated. The child may be affectionate, and may know moments of concern, but he is unable to care for his parent. He is unable to experience the parent as existing in his own right; and since the child has no clear sense of himself and what is important for his own growth, he also has no clear sense of the parent and what is involved in *his* growth. Again, consider the one-sided relationship that occurs in psychotherapy. The therapist cares for a person who is unable to care, and when the patient is able to care and could conceivably care for the therapist, the therapeutic relation comes to an end because the patient can now care for himself and be responsible for his own life.

In a meaningful friendship, caring is mutual, each cares for the other; caring becomes contagious. My caring for the other helps activate his caring for me; and similarly his caring for me helps activate my caring for him, it "strengthens" me to care for him. But to say that caring in this case is reciprocated does not imply that it is a trade—I care for you if you care for me. And this is true even if I cease to care for another simply because my caring is not reciprocated.

If we consider only examples like that of the teacher caring for the student, the psychotherapist for the patient, or the artist for the particular work of art, it might seem as if ideally all caring seeks to come to an end. This might suggest that a criterion of success in caring would be that it becomes unnecessary, that it makes itself dispensable. We speak of the student outgrowing the relationship and "standing on the shoulders" of the teacher, or of the patient becoming responsible for his own life and no longer needing the psychotherapist, or of the work of art being realized and no longer requiring the artist. But dispensing with caring is clearly not an aim in every case. Consider the relationship between parent and child. The parent helps the child to become able to care for himself without intending thereby to terminate the relationship. Similarly, we would want a mature friendship, in which each helps the other grow, to continue indefinitely.

18. Caring as a Matter of Degree Within Limits

Caring is compatible with a certain amount of blundering and lapse in interest and sensitivity to the other's needs. Recognizing honesty and devotion, for instance, as integral to caring does not imply that anything less than absolute honesty or devotion is incompatible with caring. The caring parent, the caring teacher, the caring friend, the caring writer, all have their bad days as well as their good days; there are ups and downs in caring. But, although caring may be considered to be a matter of degree in this sense, it is a matter of degree, of more or less, better or worse, within *limits.* It is a mistake to view morbid dependency as a low order of caring, or to view malevolent manipulation as "his way of caring," or to speak of caring "too much," as if overprotection were a kind of caring. These lie wholly outside the limits of caring, and to see them as examples of a low order of caring only blurs what is distinctive and important.

If I do not basically experience the other as someone or something in its own right, then, whatever else may be going on, I am *not* caring.

However much the parent provides for the child, if his primary concern is to mold a child into what he thinks the child ought to be, or if he is more interested in having the child remain fundamentally dependent on him than in the child's becoming independent and self-determining, he does not care. In such cases the child, with good reason, feels basically uncared for because he realizes he is not perceived as an individual in his own right. The writer who puts down everything interesting even though he thereby clouds the main theme and makes its development impossible, whatever else he may be doing, is not caring for the idea. Again, if I evince little desire or ability to modify my behavior in the light of what actually helps and does not help the other to grow, I am not caring.

If caring is to take place, not only are certain actions and attitudes on my part necessary, but there must also be developmental change in the other as a result of what I do; I must actually help the other grow. To determine whether I am caring, I must not only observe what I do, feel, and intend, but I must also observe whether the other is growing as a result of what I do. This, of course, does not mean that every action of mine must result directly in its growth, as if there were a one-to-one relationship, but that my actions taken as a whole must help its growth. If there is basically no growth, then, whatever else

I may be doing, I am not caring. In other words, since my actions should be guided by the direction of the other's growth and corrected by what actually goes on, if the other does, in fact, not grow, then I am not responding to its needs and I am therefore not caring.

IV

Special Features in Caring for People

19. Caring for Other People

I have tried to describe the essential features of caring considered as helping the other grow. I have not tried to differentiate among the various "others." I have not explored and clarified the important differences between a parent caring for his child and a writer caring for his "brain child," between a painter caring for his painting or a composer caring for his music, and a teacher caring for his student or a psychotherapist caring for his patient. Now, however, I shall consider caring for people specifically: first, caring for people other than myself, and second, caring for myself.

To care for another person, I must be able to understand him and his world as if I were inside it. I must be able to see, as it were, with his eyes what his world is like to him and how he sees himself. Instead of merely looking at him in a detached way from outside, as if he were a specimen, I must be able to be *with* him in his world, "going" into his world in order to sense from "inside" what life is like for him, what he is striving to be, and what he requires to grow. But only because I understand and respond to my own needs to grow can I understand his striving to grow; I can understand in another only what I can understand in myself.

In being with the other, I do not lose myself. I retain my own identity and am aware of my own reactions to him and his world. Seeing his world as it appears to him does not mean having his reactions to it, and thus I am able to help him in his world: something he is unable to do for himself. I do not have to be perplexed, for instance, to realize that he is perplexed, but because I "feel" his perplexity from the inside, I may be in a position to help him out of it. Such understanding is open to scrutiny and checking, and is a matter of my continuing development through new experiences and information.

In caring, my being *with* the other person is bound up with being *for* him as well: I am for him in his striving to grow and be himself. I experience him as existing on the "same level" as I do. I neither condescend to him (look down on him, place him beneath me) nor idolize him (look up at him, place him above me). Rather, we exist on a level of equality. Put more accurately, I am no longer aware of levels; seeing things in terms of different levels has been, so to speak, transcended. We are jointly affirmed; neither one is affirmed at the expense of the other.

What is "being with" like from the point of view of the one cared for when he realizes he is being cared for? When the other is with me, I feel I am not alone, I feel understood, not in some detached way but because I feel he knows what it is like to be me. I realize that he wants to see me as I am, not in order to pass judgment on me, but to help me. I do not have to conceal myself by trying to appear better than I am; instead, I can open myself up for him, let him get close to me, and thereby make it easier for him to help me. Realizing that he is with me helps me to see myself and my world more truly, just as someone repeating my words may give me the opportunity really to listen to myself and have the meaning of my own words come home to me more completely.

In the broad sense, "being with" characterizes the process of caring itself: in caring for another person we can be said to be basically with him in his world, in contrast to simply knowing about him from outside. This is clearly compatible with lapses in our interest and times when we are out of touch with the other person. In a narrow sense, "being with" refers to a phase within the rhythm of caring, a phase of being with the other that is followed by, and may be contrasted with, a phase of relative detachment in which we scrutinize and reflect on the experience in order to clarify our understanding and thus be more responsive to the other.

In caring for another person I encourage him, I inspire him to have the courage to be himself. My trust in him encourages him to trust himself and to be worthy of the trust. Perhaps few things are more encouraging to another than to realize that his growth evokes admiration, a spontaneous delight or joy, in the one who cares for him. He experiences my admiration as assuring him that he is not alone and that I am really for him. His awareness of my delight in his efforts to grow has a way of recalling him to himself: I help him realize and appreciate what he has done. It is as if I said to him, "Look at yourself now, see what you did, see what you can do." The opposite is responding to the growth of another with resentment, as if the other had exceeded bounds—"Just who do you think you are!" The caring mother seeing her child begin to do things for himself hugs him joyfully, and this in turn further encourages the child, and such spontaneous delight expressed in a gesture or an expression of the eyes may often carry conviction in ways that words cannot.

Admiration as spontaneous delight should not be confused with adulation. Admiration brings me closer to the one cared for; I see him as he is. In hero worship, however, I relate largely to a figment of my own imagination and am basically out of touch with the other. Also, admiration is not at the expense of yet another whom I necessarily disparage by comparison with the one I praise excessively. Adulation has nothing to do with caring.

If another person is to grow through my caring, he must trust me, for only then will he open himself to me and let me reach him. Without trust in me he will be defensive and closed. The patient must trust the psychotherapist if he is to reveal himself to him, and in this way, incidentally, the patient gives himself the chance to see and understand himself as well. The student must trust the teacher if he is to show his weaknesses and not fear being "found out," and thereby give the teacher a better idea of where to begin and what has to be done. The child must trust his parents if he is to get the help he needs. Whether I trust the one who cares for me depends in large part on his ringing true for me, and on my experiences of being actually cared for by him. On the other hand, although in my caring for another person I trust him to grow and actualize himself (to become other than he now is), he trusts me because of what I am now, because I am for him and with him in his growth.

In caring for a child I encourage him to make those decisions that are reasonably commensurate with his own powers and experiences, for one of the ways in which he grows is by developing the ability to make decisions for himself and to take responsibility for them. But the younger the child is, the less experience he has to learn from and the smaller his resources for independence, and the more I must make important decisions for him and see that they are carried out, sometimes if only for his physical safety. My firmness in making such decisions is grounded in the belief that they will help to strengthen rather than weaken his decision-making powers, and in the end will further his independence and growth. If possible, I try to help him realize that my decision was not an arbitrary exercise of authority, by explaining the reason for it and by actions which show that it was made out of concern for him.

When I care for an adult, on the other hand, I try to avoid making decisions for him. I help him make his own decisions by providing information, suggesting alternatives, and pointing out possible consequences, but all along I realize that they are his decisions to make and not my own. If I made his decisions for him, I would be condescending to him and treating him as a child; and by denying his need to take responsibility for his own life, I would be denying him as a person.

20. Caring for Myself

Just as I may be indifferent to myself, use myself as a thing, or be a stranger to myself, so I may care for myself by being responsive to my own needs to grow. I become my own guardian, so to speak, and take responsibility for my life. Caring for myself is a species of the genus "caring."

Almost all the characteristics of caring—devotion, trust, patience, humility, honesty, and the primacy of the process—apply in a straightforward way to caring for myself. However, the union with *the other* that goes with my awareness of it as existing in its own right has to be understood somewhat differently, because the other in this case is not separate from me. To care for myself, I must be able to experience myself as other (I must be able to see myself from the inside as I appear from the outside), and at the same time I must feel at one with myself rather than cut off and estranged from myself. Also, some of the ideas about caring become strained and artificial when applied literally to caring for myself. For instance, "In helping the other to grow, I grow as well" becomes "In helping myself to grow, I grow as well"; or

"In caring for another, we help him to care for himself" becomes "in caring for myself, I help myself to care for myself."

Egocentricity is morbid preoccupation with self and opaqueness to the needs of others. But there is nothing egocentric about caring for myself. First, the self-idolatry and the preoccupation with whether or not others admire me that are characteristic of egocentricity have nothing to do with helping myself to grow. In fact, the egocentric person is not fundamentally interested in himself; he avoids looking honestly at himself because he is essentially indifferent to his own needs to actualize himself. The self-complacency that often accompanies egocentricity is the converse of responding to one's own needs to grow.

Second, caring for myself takes into account my need to care for something or someone outside of myself. I can only fulfill myself by serving someone or something apart from myself, and if I am unable to care for anyone or anything separate from me, I am unable to care for myself.

Only the man who understands and appreciates what it is to grow, who understands and tries to satisfy his own needs for growth, can properly understand and appreciate growth in another; for I relate to other people in the same general way in which I relate to myself. Although caring for another person assumes that I care for myself (if I am unable to care for myself, I am unable to care for another person), the connection between caring for some other *thing* and caring for myself does not seem to be so close. The writer or the artist, it would seem, may care for his work without necessarily being able to care for himself.

V

How Caring May Order and Give Meaning to Life

21. Caring Orders Other Values Around Itself

Up to this point, the pattern of caring has been examined without considering its place within the larger context of a man's life. I would now like to examine the role of caring in a man's life, and the nature of a life that has been integrated through inclusive caring.

Caring has a way of ordering activities and values around itself; it becomes primary and other activities and values come to be secondary. When a man who has been unable to care or had no one or nothing to care for comes to care for some other, many matters previously felt to be important fade in significance, and those related to caring take on new importance. For example, if my work now gives me the opportunity to care, matters of status—whether I compare favorably with others—which previously seemed very important, become insignificant. As a caring parent, I recognize the importance of factors in my community having to do with the welfare and growth of children which I did not notice before.

As far as I am able, I promote and safeguard conditions that make my caring possible, I exclude what is incompatible with my caring and its conditions, and I subordinate what is merely irrelevant. Such ordering is not felt as an imposition from outside which denies me and closes me to life; rather, it is unforced and, like a natural unfolding, emerges from within life. It is liberating in that it opens me up more fully to life and brings me more in touch with myself and others, just as organizing my time around what I find genuinely interesting helps me to live more significantly.

People who care value caring by other people and tend to encourage and further it in others. The caring person is drawn toward other caring people, just as the interesting person (the man genuinely interested in something) is attracted toward other interesting people. If my carings are inclusive enough, they involve me deeply and fruitfully order all areas of my life. Caring then provides a center around which my activities and experiences are integrated. This results in a harmonizing of the self with the world that is deep-seated and enduring. Such harmony contrasts not only with a fragmented self in a world of unrelated experiences, but also with a self and a world upon which order has been artificially imposed. It differs from a life without inclusive commitments and from a life with inclusive commitments of a kind which in the end alienate one from oneself. This deep-seated harmony of self with the world is distinct from passively accommodating ourselves to the world or from trying to subject the world to our will.

Such inclusive ordering requires giving up certain things and activities, and may thus be said to include an element of submission. But this submission, like the voluntary submission of the craftsman to his discipline and the requirements of his materials, is basically liberating and affirming. It is like being liberated as the result of accepting some truth I have long tried to avoid; there is acceptance rather than resignation with lingering resentment, and in the end I come to realize that I would not have it otherwise. This submission entails giving up pretensions and coming to accept myself as I am; I come to see the conditions of life as they are instead of as I wish them to be. It is very different from rejecting or excluding previous ways of living by "turning over a new leaf" or "wiping the slate clean"; for instead of becoming estranged from my past because I am unable to recognize myself in it, my past, the self I have been, is now enlarged and enjoys a more expansive life.

22. *Caring Enables Me to Be "In-Place" in the World*

We are "in place" * in the world through having our lives ordered by inclusive caring. This is in contrast with being "out of place," trying to escape from the "wrong place," seeking one's "place," and indifference and insensitivity to "place." It is not as though a preexistent place were waiting for us; we are not in place as coins are in a box, but rather we both find and make our place in the same way in which the person who "finds" himself must have helped to "create" himself as well. Nor is being in-place merely some sort of reconciliation, as if we had returned to a place from which we had once been estranged. Something fundamentally new has occurred in our lives, like the change that occurs in a man's life when he comes to take full responsibility for it.

My feeling of being in-place is not entirely subjective, and it is not merely a feeling, for it expresses my actual involvements with others in

* "In place" will be used as a technical term and play an important part in what follows. To make it readily distinguishable from nontechnical uses of the term, it will be hyphenated.

the world. Place is not something I have, as if it were a possession. Rather I am in-place because of the way I relate to others. And place must be continually renewed and reaffirmed; it is not assured once and for all, for it is our response to the need of others to grow which gives us place. Again, place should not be hypostatized; it is not a thing or a fixed state. We may think of ourselves as restless, in some deep-seated sense, until we find our unique place, and of being in-place as coming to rest, but this rest is dynamic rather than static.

In-place is as much temporal as spatial, just as the present is as much spatial as temporal. Whatever else the colloquialism "He is nowhere" is supposed to mean, it does equate a person's not having a place with some basic lack in him and in his mode of living. Although for some people being in-place may be bound up with a specific location in space and time, like, for example, the artist who can only work where there are large open spaces, or the writer who cannot write away from his native soil and his native tongue, and although it necessarily presupposes some location, it is clearly not, simply or primarily, a matter of geography. Neither does being in-place imply general social acceptance. It may well result in being out of place in one's own community, just as the creative artist may be out of step with the fashions of his day. And, it has nothing to do with "knowing your proper place," where this assumes invidious class distinctions in a highly stratified society. Sometimes "knowing your proper place" may prevent you from ever being in-place.

Another rather common meaning of "place," which has nothing to do with caring and is not to be confused with being in-place, is bound up with self-aggrandizement; it means to be "in" in comparison with those people who are "out." Being in place, in this sense, requires that I experience other people as out of place and that I perceive them to be of less worth than myself. By contrast, my being in-place does not depend on my excluding others from being in-place; I have no need to compare and rank myself above others who are out of place. We may, of course, prevent another person from being in-place by interfering with the conditions necessary for his caring, but it is possible, in principle, for everyone to be in-place, for everyone to live a life significantly grounded in caring.

If caring is to be inclusive enough to enable me to be in-place, it must be rooted in my distinctive powers. It must not involve me in a peripheral way only; I must be able to make use of my particular gifts. Unless my distinctive powers are sufficiently called into play, my caring cannot significantly order my life; I am not fully engaged and I feel somewhat denied. A particular caring may have some importance for me and still not involve me enough to become central to my life. We cannot force a particular caring to be more inclusive than it is, just as we cannot force ourselves to wonder or to be genuinely interested in something. Whether some caring can become sufficiently inclusive will in the end be known only by living it, and by seeing whether, together with other carings, it can fruitfully order and integrate my life.

Beyond that, for caring to be sufficiently inclusive, a man must care for himself, because the man who is unresponsive to his own needs to grow can never be in-place. And, as we have seen, caring for myself means at the least to be responsive to my fundamental need to care for others apart from myself.

Further, my carings must be compatible, in some kind of harmony with one another, if they are to be inclusive enough to enable me to be in-place. My life cannot be harmoniously ordered if, for example, there is a basic incompatibility between caring in my work and caring for my family, or between caring for myself and caring for some particular other person. Harmony can tolerate occasional conflicts in priority; there will be times when this caring rather than that comes first. But such conflicts are evened out and resolved over the long run, just as every significant friendship and every happy family resolves similar conflicts in due course. How many carings are required to order our lives fruitfully is, of course, an individual matter, but the number is always small, for we cannot really be devoted to many things at the same time. We may have many acquaintances, but we deceive ourselves if we think it is possible to have many friends; at the very least, there is simply not enough time.

23. Caring for My "Appropriate" Others

To simplify matters, I will call those I care for inclusively enough to enable me to be in-place, "appropriate others." The word "appropriate" is not, of course, the important thing; "fulfilling" might have been used instead. My appropriate others complement me, they enable me to be complete, somewhat as playing music enables the musician to be himself. We sometimes say, "I was incomplete until I found interesting work," or "Until I met you, I felt incomplete," or "I finally felt complete when I had my own family." This sense of completeness does not mean the end of growth, as if one were now somehow finished; rather, it goes with being in the best position for further growth.

The relationship between my appropriate others and myself is not external, like that of a chair and a table. Instead, I experience them as an extension of myself and I identify with their growth. But neither is the relationship parasitic; they are part of me in a way that affirms us both.

And although in all cases of caring I experience the other as, in a sense, part of me, the experience is much more pronounced when the other is an appropriate other and is felt to complete me. Indifference to such an other turns out to be indifference to myself also, and in time results in loss of place. To be in-place then is living that is centered and integrated by my caring for my appropriate others, one of whom, to repeat, must always be myself.

My appropriate others, speaking now simply of those separate from me, are not ready-made and waiting for me. They must have developed in relation to me to the point where, in conjunction with other carings, they have become a center around which my life can be significantly ordered. And in helping them grow to this point I myself am transformed: in finding and developing my appropriate others I find and create myself. In much the same way ideals can become active and important in my life rather than merely professed. I do not find them ready-made; they have to be developed and clarified through my acting on them; they must be made actual, extended, and safeguarded. I must, as it were, become the ideals in order fully to actualize them; and by making them part of myself, I too am transformed. Just as the other must grow until it becomes central to my life, so I must grow until I can have such inclusive others. I must become, for example, the kind of person who can significantly devote himself to something for its own sake.

Consider the mutual development of the other and myself where a "brain child," say, a seminal philosophical idea, has become an appropriate other. Through my efforts it has grown from something amorphous and peripheral to something more definite and central to my life. I have made it my own, not in the sense that I own or possess it, since I experience it as existing in its own right, but because I have assimilated it and committed myself to it. I have been transformed in the course of helping it grow to that point of profound involvement with me where it helps to order my life fruitfully. Or consider the mutual development of the other and myself where a friend has become an appropriate other. Together with the changes that have taken place in him and in my relations to him, I have grown significantly as well. At the very least, by coming to experience the world through his eyes as well as my own, my self has been enlarged and my understanding broadened.

Whether a particular other is an appropriate other may depend on its being seen from a short- or long-term point of view. From a short-term point of view, a specific book that a man is writing may be an appropriate other for him, but from a long-term point of view, the appropriate other may be, instead, certain ideas or ideals expressed in all his writings. His caring for a particular book could then also be understood as a way of caring for them. If appropriate others were thought of from a short-term point of view only, it would imply that when a writer finishes one book and begins another, he necessarily changes his appropriate other. This would make it impossible to recognize and appreciate objects of continuing devotion. Interpreted from a long-term point of view, however, his appropriate other might well remain basically the same, and this would be equally true if he stopped writing and began teaching as a way of caring for his ideas.

I am *on call* for my appropriate others. This does not simply mean I am available in the sense of being open and receptive, but corresponds to the way the person "off duty" may be reached and called in when he is needed. The man who cares for his appropriate others aspires to be always available to them when they really need him: the caring parent can be called away from something else to return to his child; the caring doctor can be reached by his patient; the caring artist is at the call of his work of art. Wherever I am, whatever I may be doing, I am subject to being called in by my appropriate others. In this sense, I may be said to be always on hand.

When I care for appropriate others, devotion, trust, hope, and courage stand out all the more clearly. For the more essential the other is to me, the more fully evolved is the pattern of caring, and the more discernible are its different ingredients.

24. Living the Meaning of My Life

Through finding and helping to develop my appropriate others, I discover and create the meaning of my life. And in caring for my appropriate others, in being in-place, I *live* the meaning of my life. This meaning is not felt to be external to my life and tacked onto it; there are a rightness and necessity about it that are rooted in my being: the meaning is acknowledged to be my own. I realize the point of my life: what specifically requires me, what it is I am to serve. The specific need certain others have for me, if they are to grow, makes it possible for me to live the meaning of my life. Correspondingly, unless I discover what basically requires me, it is not possible for me to find this meaning.

No one else can give me the meaning of my life; it is something I alone can make. The meaning is not something predetermined which simply unfolds; I help both to create it and to discover it, and this is a continuing process, not a once-and-for-all.

To live the meaning of my life is to be in-place by virtue of caring for my appropriate others. Clearly, such a life is not necessarily one of maximum pleasure and comfort. It may include much hardship and sorrow, and it need not be the richest cultural life. But in a distinctive sense it is my own life, one rooted in my own being and not foreign to me. No significant gap exists between how I live my life and who I really am or how I should be living. I do not feel that I should be somewhere else doing something else; the way I live is reasonably commensurate with my powers and limitations. If there is any sense to the question, "What is *the* meaning of life?" I alone can answer it for myself by finding and living the meaning of my own life. To experience meaning in my life because of some purpose, or to experience meaningful times, or to have other people view my life as having meaning, does not necessarily imply that I live the meaning of my life.

I have a calling by virtue of responding to the call of my appropriate others. This does not mean that these others have been predetermined for me. It is as if *I* had been chosen or "singled out" to care for them; it is *my* unique task. But to be thus singled out does not imply disparagement of anyone else: I am not chosen at someone else's expense. Although I experience a greater sense of my own individuality, and I am set apart in the sense that this is *my* task, I am not cut off from my fellows; if anything, I am closer to them because I can more easily understand and appreciate how they can be singled out for *their* unique task. And being chosen in this sense does not mean that I passively receive what happens to me from outside. By finding and helping to develop my appropriate others, I have taken a hand, so to speak, in singling myself out.

When we speak of "having direction" in our lives, we mean roughly that something (an aim, a goal, an interest, a person, an ideal) attracts us enough to move us to action on its behalf and is inclusive enough so that focusing upon it coordinates our activities and provides our lives with a measure of continuity. But this kind of direction may be found in ways of living that are very different in content and worth. It may accompany self-understanding and taking responsibility for one's own life; and it may also accompany lack of self-understanding and avoiding or abdicating such responsibility, like the direction that comes from blind adherence to a dictator. Simply having direction does not imply that I live the meaning of my life.

Since I live the meaning of my life by caring for my appropriate others, I could live differently, I could have different appropriate others (with the exception of myself, of course), and still live the meaning of my life. But to what extent I could have significantly different appropriate others depends on my particular powers and limitations, the general situation I find myself in, and the resources available to me. For some, particularly in the area of work, the possible others that a man could inclusively care for may be very limited, and if certain areas are denied him, it may be impossible for him to find and live the meaning of his life.

Above all, it would be a mistake to think that living the meaning of my life is primary if this suggests something over and above caring for my appropriate others. In that case caring would be viewed as a means for achieving something felt to be more important. But I do not care in order to live the meaning of my life. Instead, living a life centered around caring for my appropriate others *is* living the meaning of my life. And it is only because the others I care for are primary for me that I am able to live the meaning of my life.

VI

Major Characteristics of a Life Ordered Through Caring

25. Basic Certainty

There is a stability in being in-place that is enduring and is connected with one's way of living in general; it is not something temporary or related only to this or that specific situation. I am steadied and centered in my living, and the discordant experiences of day-to-day living, experienced against a settled background, are more easily assimilated. Such general stability can withstand much stress and, like devotion, is strengthened by overcoming difficulties. It resembles the grounding and firmness that result from having a unifying purpose or aim, or from overcoming the need to pretend to be something we are not, or from having a good sense of our own worth. This stability is the antithesis of the deep-seated uncertainty as to who we are and what we are about that goes with ambivalence and drift, and it differs radically from the certainty of the fanatic, which exists against a background of fundamental uncertainty. It is as if we have come to sleep on the floor and therefore no longer fear falling out of bed.

To speak of this stability as basic certainty has nothing to do with claims of possessing the truth or certain knowledge. Nor is it to be confused with the feeling of certainty that comes from clinging to some authority or belief. Basic certainty is more like being rooted in the world than like clinging to a rock. Thanks to it we are open and accessible, whereas clinging closes us to experience and makes it impossible for us to be sure about whom or what we cling to. And basic certainty has nothing to do with the certainty we try to gain by avoiding confronting and being honest with ourselves, in the way we sometimes tend to disregard negative evidence in matters in which we have some significant personal stake.

Basic certainty requires outgrowing the need to feel certain, to have absolute guarantees as to what is or will be. Instead, if we think of basic certainty as including deep-seated security, it also includes being vulnerable and giving up the preoccupation with trying to be secure.

The stability of a life centered in caring is a continuing matter, not something static and acquired once and for all. And basic certainty does not exist inside a man as a stoical resolve to remain unperturbed in a perilous world. In fact, because it concerns the particular way a man is related to others, it is only in a derivative and abstract sense that we can attribute it to any individual. I am sustained and supported by entering more significantly into life; if living the meaning of my life is interfered with, basic certainty is interfered with.

The experience of *belonging* that stems from being needed by my appropriate others helps ground me; it is an ingredient of basic certainty. I belong because my appropriate others need me, because I have been entrusted, as it were, with the being of these others. The man who is not needed by someone or something does not belong and lives like a leaf blown about in the wind. I have a need to be needed, and the need of others for me goes hand in hand with my need for them. Belonging, in this sense, goes with my own actualization, and is very different from morbid dependency, in which I lose my own integrity.

The union between *inner and outer* is another ingredient of basic certainty. When I am in-place, there is a reasonable convergence between my professed values and how I actually live, between how I think and how I actually live, between how I see my behavior from inside and how it looks to others from outside. My living testifies to what my values are. To the extent that there is significant division between inner and outer, I am not wholly in my actions; I am divided within myself, and in the end must be uncertain who I am and what I am about.

In basic certainty we discover also the *clarity* that results from the elimination of much clutter. A simplification takes place in my life through recreating it around caring. Much that is incompatible and irrelevant to my caring is eliminated, and I achieve a fundamental clarity as to who I am and what I am about. This is the kind of clarity, for example, that comes with ceasing to be preoccupied with matters of status and becoming absorbed instead in matters I find genuinely interesting, or with recognizing what I really want and what means are required to attain it, or with overcoming the need to pretend to be what I am not. With the elimination of clutter, living becomes less complicated; important connections between events are more easily seen, and the significance of experiences more readily comes home to us. There is a greater directness in life, and what is of real importance stands out more clearly. Without such clarity deep-seated stability is impossible.

Clutter not only interferes with creating new possibilities, but, because it satiates and dulls me, it interferes with focusing on what is in front of me and seeing the possibilities that already exist. The writer, for instance, is sometimes immobilized by trying to work with too many ideas at the same time. The ideas get in each other's way, and it is only when he limits them to what is directly relevant to the point at hand that there is enough space and time "between" them so that they reverberate with meaning and he can see vital connections between them. While clutter shrinks my "living room" and interferes with acting freely, simplification, grounded in responding to what is basic to me, expands my living room and is liberating. The resulting economy makes for depth rather than shallowness in living. Space and time seem more spread out, making possible a greater play of the imagination and the development of new possibilities. I can see what I am doing, I can focus more easily on something and develop my relationship to it. If clutter is thought of as a kind of interfering noise, like static, there is a certain quieting that goes with simplification and basic certainty.

Just as in the case of living the meaning of my life, to assume that achieving basic stability is primary and that caring is simply a means toward this end is to be fundamentally confused about both. When we use our others in this way, we interfere with caring and make basic certainty impossible. It is not that we care in order to achieve basic certainty, but that a life centered in caring has this stability. Ultimately, this kind of stability can only come through forgetting about trying to achieve it and concerning ourselves instead with the others that need us.

26. The Process of Life Is Enough

The process of living is experienced as enough in itself when I live the meaning of my life. This does not imply perfection, however we may think of perfection. When we admit that a friend, a conversation, a musical performance, or a book is not perfect but is "good enough," it is not that we believe improvement impossible, but that improvement would not fundamentally change matters. Life is felt to be enough in the living, and what I want is simply the opportunity to live this life.

Clearly, being "good enough" does not imply the maximization of pleasure over pain. Caring is not always agreeable; it is sometimes frustrating and rarely easy. And "good enough" has nothing to do with stagnation or complacency, for the process is good enough only because of the creative way in which I live. Also, to speak of present living as enough is compatible with the unfinished character of living; man as growing and creating is always unfinished and in the making. In fact, in the fullness of living the meaning of my life, I am more than ever aware of the unfinished character of living.

Perhaps several examples in which we experience ourselves as fundamentally denied and the process of living as not enough may help to suggest, by contrast, what it is like when it is enough. Living is felt not to be enough when we do not utilize our distinctive powers (when the writer is prevented from writing and the nurse is prevented from nursing). At such times we do not experience ourselves as efficacious and we feel outside life, as if life were passing us by, because we are not significantly engaged in it. Or we feel this lack when we are constantly hurried and feel that we need more time in order to make contact with anything. Then the tempo of life prevents our undergoing our experiences fully enough to assimilate them. Also, the process of living is felt not to be enough when we are basically pretentious and present ourselves as being something we are not; we feel that what we are now is never quite enough, and we would rather be treated as we think we will be rather than as we are. And, certainly, present living is felt not to be enough when we always seek rather than live the meaning of our life, when we feel there is something to be reached which constantly eludes us like the carrot ahead of our nose, and the present is simply a necessary evil which stands in our way. On the other hand, present living is enough when I live the meaning of my life. I do not experience a need to get to life, as if it were something beyond or outside present living. And when present living is enough, I experience myself as being enough.

There is a density about experience when we are in-place that is the antithesis of the sense of impoverishment characteristic of pervasive boredom, when there is nothing to explore, and novelty is lacking, and nothing grows. This density shows itself in the inexhaustible character of caring, for caring is not like a fixed quantity that can be used up; rather, it is renewed and developed through exercise. This fullness also shows itself in the growth of our appropriate others through our caring for them. Existence opens up in depth; it is like inquiring into interesting subject matter and finding the materials richer instead of leaner the deeper we go. Intimations of unsounded depths color our experience when the process of living is felt to be enough.

27. *Intelligibility and Unfathomability*

A pervasive intelligibility comes into my life when I live the meaning of my life, but not in the sense in which scientific explanations or the ability to predict and control make phenomena intelligible. Nor is it what we feel in the presence of anything familiar. Instead, it consists in understanding what is relevant to my life, what it is that I live for, who I am and what I am about in actual day-to-day living, not in the abstract. By contrast, the man who continually seeks the meaning of his life, who is confused about what is or would be relevant for his growth and is therefore unsure who he is, lives in a world that does not quite make sense. And similarly the world lacks intelligibility for the man who drifts with no sense of purpose, or the man whose life has a direction which has been forced on it and is alien to him. The intelligibility I am trying to suggest goes with feeling that we belong and are uniquely needed by something or someone, in contrast to the disquiet that comes with not quite fitting in anywhere and with continued and sometimes desperate attempts to find our place. It is bound up with my feeling understood and cared for by myself, and is at the opposite pole to the unintelligibility I feel when I am out of touch with myself, am unresponsive to myself and my primary needs, and am unable to learn significantly from my past.

Such intelligibility is not a once-and-for-all thing, but is a continuing function that goes with caring for my appropriate others. Thus I play an active part in making my world intelligible. And it is not simply that I have an understanding of who *I* am and what *I* am about. The more deeply I understand the central role of caring in my own life, the more I realize it to be central to the human condition. My world becomes intelligible for me through caring and being cared for, or, put differently, as I become responsible for the growth and actualization of others. In the sense in which intelligibility means being at home in the world, we are ultimately at home not through dominating or explaining or appreciating things, but through caring and being cared for.

Such intelligibility does not diminish or do away with wonder. It is, instead, conducive to wonder because it makes me more open to myself and the world. By "wonder," I do not mean puzzlement or perplexity, that is, something to resolve, but something to savor, like the wonder in looking up at the stars at night. Sometimes we may even feel that we truly perceive an object only to the extent that we experience it with wonder. Insofar as man grows and lives creatively, wonder is naturally present in his life; but if his growth is seriously blocked, he becomes closed to life and wonder becomes almost impossible.

Similarly, intelligibility does not do away with the unfathomable character of existence, but rather enables us to be more aware of it. The unfathomable character of existence is not a matter of ignorance to be resolved, it is not something to overcome by knowing more or having some special knowledge. Instead, like wonder, it is something to undergo, to realize, and to appreciate. I am not simply speaking about the mystery of coming into being and passing out of being, or the strange sense that

I was not here at the beginning and will not be here at the end. I am speaking, rather, about the mystery of existence itself, the mystery and amazement that anything exists at all.

This awareness of unfathomability is not something to fear and flee from, but to realize deeply. Unlike the experience of the uncanny, it does not separate me from other people; it brings me closer by making me more aware that whatever our powers or limitations, whatever our possessions or lack of possessions, we are all in the same boat. This is not a leveling down that does away with differences. On the contrary, it makes for a greater appreciation of the uniqueness of others and of myself. I realize more deeply my own insignificance, as if I were a brief flame in an endless darkness, and I am also more aware of my incomparable worth, a preciousness that is somehow bound up with being a once-and-for-all, never to be repeated.

As we have said, a simplification in living comes with being in-place which makes for growth and meaning rather than shallowness in living. By eliminating much that is incompatible and irrelevant to my carings, what is important to me stands out more clearly and I can be more aware of who I am and what I am about. This elimination of clutter also enables me to be more readily aware of the unfathomable character of life, in much the same way that the elimination of noise and distraction enables me to be more aware of myself and the stillness surrounding me. In this sense, one may say that a greater appreciation of the unfathomable character of existence accompanies the intelligibility of being in-place.

28. Autonomy

Autonomy can be significantly equated with living the meaning of my life, for, within certain limits set by the social and physical conditions under which I live, I go my own way. This is not true of living a life that is foreign to me, that is not felt to be my own, in which I live according to values which are not rooted in my own experience and are indifferent to my needs to grow and actualize myself. Of course, "a life of my own" does not mean that my life is a possession I can use and manipulate in any way I please, for if I treat myself as a possession, I become estranged from myself and my life becomes foreign to me. In order to live "my own life" I must make it my own through caring and taking responsibility for it, just as I must act on an ideal and help to actualize it if I am to make it my own. I am not autonomous to begin with; autonomy is an achievement like maturity or the growth of a significant friendship.

Autonomy does not mean being detached and

without strong ties; that would imply that attachments and strong ties necessarily tie me down and enslave me. Again, autonomy does not mean being self-enclosed and "free as a bird." On the contrary, I am autonomous because of my devotion to others and my dependence on them, when dependence is the *kind* that liberates both me and my others. Consider, for instance, how my powers are liberated when a superficial relationship develops into a significant friendship, or when I fully commit myself to some interesting work. The teacher cannot be himself without deep commitment to his students, and the musician cannot be himself without deep commitment to his music. Such commitments to others are essential if one is to be himself. It is because of this dependence on certain others that I live the meaning of my life, that I can "live my own life." Neither is autonomy the absence of others' dependence on me; rather, it is made possible because I am needed by my appropriate others. The artist's freedom is grounded in his need for his work, in his commitment to his work, and in his being needed by his work. I am autonomous because in helping my appropriate others to grow I also grow. Autonomy is not a matter of my own actualization if this be taken as separate from the actualization of others. Sometimes people say in despair, "If only I could care for someone or something!" implying they are not free because they do not feel needed, but would be free if they were, if they had something to serve.

Autonomy does not mean doing anything I please; in caring for my appropriate others I do not act arbitrarily. How and what I do are significantly determined by what is required of me by my others if they are to grow and be actualized, and by my own needs to grow. Although my direction is largely determined by the growth of others, I experience myself as the initiator of my acts and as responsible for my own life, and not simply as acted on and controlled from outside. Direction emerges within my life, instead of being something predetermined or forced on me from outside. Autonomy is the opposite of both arbitrary behavior (doing simply as I please) and behavior controlled by what is basically foreign to me.

The present is liberating when I am in-place; it is fertile and things grow. It contains things that I cherish and am devoted to. Past and future, instead of being fixed and laid out for me, have an unknown and promising dimension. My awareness of the present is deepened by reflecting on the past, on what was and what could have been, and reflecting on the future opens me to rich possibilities that move me to act on their behalf. The liberation of my distinctive powers goes hand in hand with a deepening and widening of the present. Since I identify with the growth and well-being of others outside of me, there is an expansion of self.

Consider, by contrast, pervasive boredom as an attitude toward life in general rather than toward a particular situation or activity. In pervasive boredom the present is like a desert in which nothing grows and nothing is precious. I am unable to animate anything or be animated by others. Reflecting on the past or future only aggravates and reinforces the emptiness and futility of the present. Because I am unable to make vital contact with what is around me, there is a contraction of self. My life lacks purpose; nothing moves me to devotion on its behalf, or, put differently, there is nothing that I feel worth serving. Everything is surface; there is nothing to explore which might open me to more vital ways of living. Life lacks promise: whatever happens, whatever I do, life will continue with the same dull monotony. My distinctive powers are immobilized and I am unfree. But through being in-place I am autonomous because I live in a liberating present.

Autonomy assumes self-understanding; in the end, without such understanding I get in my own way and go around in circles. In being in-place I understand myself in important ways. First, there is the understanding necessarily present in my actually caring for myself: understanding who I am, what I am striving for, what my needs are, and what is required to satisfy these needs. Clearly, I can never understand myself once and for all, for self-understanding, like caring itself, is a continuing matter. And, like caring, it is a matter of more or less within limits. Second, there is a more inclusive self-understanding found in being in-place, which includes knowing what I am to serve, what is required of me, what complements me. I show such understanding by living it, by caring for my appropriate others, and this too is a continuing matter and not achieved once and for all. Self-understanding in this inclusive sense is finding and living the meaning of my life.

In being in-place I am both significantly immersed in life and, at the same time, *free* of certain ways of living, widespread in our society, which are hostile to growth. I am free of experiencing life as a *race* in which I am concerned with how I compare with others: whether they are ahead or behind me; whether I am catching up to them, maintaining my distance from them, or falling behind them. Since I am not in the race, I am not caught up in the humiliation and the vindictiveness that are so much a part of the race. Also, I am free of experiencing life as a *market place* in which I see myself and others as commodities to be sold, and try to make myself into the package that happens to be in demand at a particular time. I do not experience the impotence that comes with having my sense of identity depend basically on the opinions of others instead of on the use of my own powers. Also, I am free of the *discontinuity* and chaos of experiencing life as a mere succession of disconnected events, each unrelated to what went before and what is to come. When life is felt to be made up of unrelated fragments, there can be no appreciation of growing and maturing, and of working out or through something.

29. Faith

Faith is present in both a narrow and a broad sense in being in-place. In a narrow sense, it is an ingredient of living the meaning of my life, as when we speak of having faith *in* someone or something. I have faith, for example, in my powers to care for myself, where such care includes being responsive to my need to care for others. Or, more generally, I have faith in my ability to learn from experience and to be attracted by growth and what is life-furthering in experience, as a plant is attracted to sunlight. Such faith dissipates the fear of possible self-betrayal through a sick conformity in which, for the approval and security of the herd, I become indifferent to my own needs and the needs of others to grow. Faith in myself is neither blind nor irrational; it is warranted by my experiences of caring and being cared for, just as my faith in another person's concern for my growth (that he will not betray my need to be myself) is warranted by my experience of his caring for me.

There is also a broader sense in which faith can be equated with being in-place itself. This is faith as my general orientation: I stand forth in the world and commit and expose myself. As a man can be said to be in his words or in his acts, I am in my life.

The contrast between faith and lack of faith, in this broader meaning, may be suggested by the difference between the man who takes responsibility for his life and the man who avoids this responsibility and wants others to take it over. Faith as a way of being, as a basic trust in life, goes with confidence in going into the unknown in the course of realizing ourselves and caring for our others. It is the antithesis of closing ourselves off through fear of the unknown; instead of avoiding life, we are more accessible to it. Such faith is, of course, not incompatible with lack of faith in particular people or particular ideals based on our experience of them.

Since we speak of faith in the narrow sense as *faith in*, perhaps we could speak of faith in the broader sense as living *in faith*. Then we could say that when a man cares for his appropriate others, when certain inclusive others need him if they are to grow, he is in faith, and, by being in faith, he is rooted in the world. Such faith comes from entering more significantly into life. It is like the confidence and security that come only with the fullest use of our distinctive talents. For the man who is in faith, the world is intelligible, but it is an intelligibility which goes with a more acute awareness of the unfathomable character of existence.

30. Gratitude

Gratitude is a natural expression of being in-place. I am grateful for living the meaning of my life (for having appropriate others that need me and for being able to care for them) and, more generally, for life itself. I am thankful for the opportunity and the capacity to give of myself. It is because I give that I receive, which, of course, does not mean that I give in order to receive. I cannot be grateful for what I believe another was forced to give me, and the conviction that "it was due me" is equally incompatible with gratitude. Also, it is not that a claim against life has been satisfied, that I must be allowed to live the meaning of my life. If I think of being in-place as something I can demand of life, I cannot feel gratitude.

By realizing that what I receive did not have to be given me and that it was not something I could demand, I appreciate how profoundly dependent I am on numerous factors over which I have little or no control. But recognizing such dependence is not debilitating or degrading; it is, on the contrary, liberating and joyful. I feel close to what I am dependent on just as the true craftsman feels affection for his tools and his materials. Instead of grudgingly acknowledging my dependence on others, I feel gladness in realizing my debt to them and the ties that bind us together; it is like the gladness found in significant sharing. Such appreciation of dependence goes with independence in living the meaning of my life. Only that man who "helps" others as a way of manipulating them will experience receiving help as imprisoning; he cannot take in because of the fear of being taken in.

Gratitude remains incomplete until I have expressed my thanks for what I have received. But how do I thank and whom do I thank? Gratitude for being in-place makes me experience people and things as more precious, and I become more responsive to them and their need for me; gratitude further activates me to care for my appropriate others. **Caring becomes my way of thanking for what I have received; I thank by caring all the more for my appropriate others and the conditions of their existence.** It is, in some ways, like showing appreciation for a gift by using it fully. Just as a man may be said to love or revere life, or to despise life, so I may be said to thank life for what I have received; or, if nature is understood broadly as the source of all there is, I may be said to thank nature. It is as if I am cared for by being in-place, and I want to reciprocate. But I cannot thank (care for) life in general. I can only thank life by caring for this or that instance of it.

Man finds himself by finding his place, and he finds his place by finding appropriate others that need his care and that he needs to care for. Through caring and being cared for man experiences himself as part of nature; we are closest to a person or an idea when we help it grow. There is a rock-bottom quality about living the meaning of my life that goes, oddly enough, with greater awareness of life's inexhaustible depths; it is as if life is ordinary and "nothing special" when it is most extraordinary. And although we find a deep-seated intelligibility in life, the last word is with the unfathomable character of existence which, like a pedal point in a piece of music, pervades and colors life.

World Perspectives

What This Series Means

It is the thesis of World Perspectives *that man is in the process of developing a new consciousness which, in spite of his apparent spiritual and moral captivity, can eventually lift the human race above and beyond the fear, ignorance, and isolation which beset it today. It is to this nascent consciousness, to this concept of man born out of a universe perceived through a fresh vision of reality, that* World Perspectives *is dedicated.*

Man has entered a new era of evolutionary history, one in which rapid change is a dominant consequence. He is contending with a fundamental change, since he has intervened in the evolutionary process. He must now better appreciate this fact and then develop the wisdom to direct the process toward his fulfillment rather than toward his destruction. As he learns to apply his understanding of the physical world for practical purposes, he is, in reality, extending his innate capacity and augmenting his ability and his need to communicate as well as his ability to think and to create. And as a result, he is substituting a goal-directed evolutionary process in his struggle against environmental hardship for the slow, but effective, biological evolution which produced modern man through mutation and natural selection. By intelligent intervention in the evolutionary process man has greatly accelerated and greatly expanded the range of his possibilities. But he has not changed the basic fact that it remains a trial and error process, with the danger of taking paths that lead to sterility of mind and heart, moral apathy and intellectual inertia; and even producing social dinosaurs unfit to live in an evolving world.

Only those spiritual and intellectual leaders of our epoch who have a paternity in this extension of man's horizons are invited to participate in this Series: those who are aware of the truth that beyond the divisiveness among men there exists a primordial unitive power since we are all bound together by a common humanity more fundamental than any unity of dogma; those who recognize that the centrifugal force which has scattered and atomized mankind must be replaced by an integrating structure and process capable of bestowing meaning and purpose on existence; those who realize that science itself, when not inhibited by the limitations of its own methodology, when chastened and humbled, commits man to an indeterminate range of yet undreamed consequences that may flow from it.

Virtually all of our disciplines have relied on conceptions which are now incompatible with the Cartesian axiom, and with the static world view we once derived from it. For underlying the new ideas, including those of modern physics, is a unifying order, but it is not causality; it is purpose, and not the purpose of the universe and of man but the purpose in *the universe and* in *man. In other words, we seem to inhabit a world of dynamic process and structure. Therefore we need a calculus of potentiality rather than one of probability, a dialectic of polarity, one in which unity and diversity are defined as simultaneous and necessary poles of the same essence.*

Our situation is new. No civilization has previously had to face the challenge of scientific specialization; and our response must be new. Thus this Series is committed to ensure that the spiritual and moral needs of man as a human being and the scientific and intellectual resources at his command for life *may be brought into a productive, meaningful, and creative harmony.*

This Series endeavors to point to a reality of which scientific theory has revealed only one aspect. It is the commitment to this reality that lends universal intent to a scientist's most original and solitary thought. By acknowledging this frankly we shall restore science to the great family of human aspirations by which men hope to fulfill themselves in the world community as thinking and sentient beings. For our problem is to discover a principle of differentiation and yet relationship lucid enough to justify and to purify scientific, philosophic and all other knowledge, both cognitive and intuitive, by accepting their interdependence. This is the crisis in consciousness made articulate through the crisis in science. This is the new awakening.

Each volume presents the thought and belief of its author and points to the way in which religion, philosophy, art, science, economics, politics and history may constitute that form of human activity which takes the fullest and most precise account of variousness, possibility, complexity and difficulty. Thus World Perspectives *endeavors to define that ecumenical power of the mind and heart which enables man through his mysterious greatness to re-create his life.*

This Series is committed to a re-examination of all those sides of human endeavor which the specialist was taught to believe he could safely leave aside. It attempts to show the structural kinship between subject and object; the indwelling of the one in the other. It interprets present and past events impinging on human life in our growing World Age and envisages what man may yet attain when summoned by an unbending inner necessity to the quest of what is most exalted in him. Its purpose is to offer new vistas in terms of world and human development while refusing to betray the intimate correlation between universality and individuality, dynamics and form, freedom and destiny. Each author deals with the increasing realization that spirit and nature are not separate and apart; that intuition and reason must regain their importance as the means of perceiving and fusing inner being with outer reality.

World Perspectives *endeavors to show that the conception of wholeness, unity, organism is a higher and more concrete conception than that of matter and energy. Thus an enlarged meaning of life, of biology, not as it is revealed in the test tube of the laboratory but as it is experienced within the organism of life itself, is attempted in this Series. For the principle of life consists in the tension which connects spirit with the realm of matter, symbiotically joined. The element of life is dominant in the very texture of nature, thus rendering life, biology, a transempirical science. The laws of life have their origin beyond their mere physical manifestations and compel us to consider their spiritual source. In fact, the widening of the conceptual framework has not only served to restore order within the respective branches of knowl-*

edge, but has also disclosed analogies in man's position regarding the analysis and synthesis of experience in apparently separated domains of knowledge, suggesting the possibility of an ever more embracing objective description of the meaning of life.

Knowledge, it is shown in these books, no longer consists in a manipulation of man and nature as opposite forces, nor in the reduction of data to mere statistical order, but is a means of liberating mankind from the destructive power of fear, pointing the way toward the goal of the rehabilitation of the human will and the rebirth of faith and confidence in the human person. The works published also endeavor to reveal that the cry for patterns, systems and authorities is growing less insistent as the desire grows stronger in both East and West for the recovery of a dignity, integrity and self-realization which are the inalienable rights of man who may now guide change by means of conscious purpose in the light of rational experience.

The volumes in this Series endeavor to demonstrate that only in a society in which awareness of the problems of science exists can its discoveries start great waves of change in human culture, and in such a manner that these discoveries may deepen and not erode the sense of universal human community. The differences in the disciplines, their epistemological exclusiveness, the variety of historical experiences, the differences of traditions, of cultures, of languages, of the arts, should be protected and preserved. But the interrelationship and unity of the whole should at the same time be accepted.

The authors of World Perspectives *are of course aware that the ultimate answers to the hopes and fears which pervade modern society rest on the moral fiber of man, and on the wisdom and responsibility of those who promote the course of its development. But moral decisions cannot dispense with an insight into the interplay of the objective elements which offer and limit the choices made. Therefore an understanding of what the issues are, though not a sufficient condition, is a necessary prerequisite for directing action toward constructive solutions.*

Other vital questions explored relate to problems of international understanding as well as to problems dealing with prejudice and the resultant tensions and antagonisms. The growing perception and responsibility of our World Age point to the new reality that the individual person and the collective person supplement and integrate each other; that the thrall of totalitarianism of both left and right has been shaken in the universal desire to recapture the authority of truth and human totality. Mankind can finally place its trust not in a proletarian authoritarianism, not in a secularized humanism, both of which have betrayed the spiritual property right of history, but in a sacramental brotherhood and in the unity of knowledge. This new consciousness has created a widening of human horizons beyond every parochialism, and a revolution in human thought comparable to the basic assumption, among the ancient Greeks, of the sovereignty of reason; corresponding to the great effulgence of the moral conscience articulated by the Hebrew prophets; analogous to the fundamental assertions of Christianity; or to the beginning of the new scientific era, the era of the science of dynamics, the experimental foundations of which were laid by Galileo in the Renaissance.

An important effort of this Series is to reexamine the contradictory meanings and applications which are given today to such terms as democracy, freedom, justice, love, peace, brotherhood and God. The purpose of such inquiries is to clear the way for the foundation of a genuine world *history not in terms of nation or race or culture but in terms of man in relation to God, to himself, his fellow man and the universe, that reach beyond immediate self-interest. For the meaning of the World Age consists in respecting man's hopes and dreams which lead to a deeper understanding of the basic values of all peoples.*

World Perspectives *is planned to gain insight into the meaning of man, who not only is determined by history but who also determines history. History is to be understood as concerned not only with the life of man on this planet but as including also such cosmic influences as interpenetrate our human world. This generation is discovering that history does not conform to the social optimism of modern civilization and that the organization of human communities and the establishment of freedom and peace are not only intellectual achievements but spiritual and moral achievements as well, demanding a cherishing of the wholeness of human personality, the "unmediated wholeness of feeling and thought," and constituting a never-ending challenge to man, emerging from the abyss of meaninglessness and suffering, to be renewed and replenished in the totality of his life.*

Justice itself, which has been "in a state of pilgrimage and crucifixion" and now is being slowly liberated from the grip of social and political demonologies in the East as well as in the West, begins to question its own premises. The modern revolutionary movements which have challenged the sacred institutions of society by protecting social injustice in the name of social justice are here examined and re-evaluated.

In the light of this, we have no choice but to admit that the unfreedom *against which freedom is measured must be retained with it, namely, that the aspect of truth out of which the night view appears to emerge, the darkness of our time, is as little abandonable as is man's subjective advance. Thus the two sources of man's consciousness are inseparable, not as dead but as living and complementary, an aspect of that "principle of complementarity" through which Niels Bohr has sought to unite the quantum and the wave, both of which constitute the very fabric of life's radiant energy.*

There is in mankind today a counterforce to the sterility and danger of a quantitative, anonymous mass culture; a new, if sometimes imperceptible, spiritual sense of convergence toward human and world unity on the basis of the sacredness of each human person and respect for the plurality of cultures. There is a growing awareness that equality may not be evaluated in mere numerical terms but is proportionate and analogical in its reality. For when equality is equated with interchangeability, individuality is negated and the human person extinguished.

We stand at the brink of an age of a world in which human life presses forward to actualize new forms. The false separation of man and nature, of time and space, of freedom and security, is acknowledged, and we are faced with a new vision of man in his organic unity and of history offering a richness and diversity of quality and majesty of scope hitherto unprecedented. In relating the accumulated wisdom of man's spirit to the new reality of the World Age, in articulating its thought and belief, World Perspectives *seeks to encourage a renaissance of hope in society and of pride in man's decision as to what his destiny will be.*

World Perspectives *is committed to the recognition that all great changes are preceded by a vigorous intellectual re-evaluation and reorganization. Our authors are aware that the sin of* hubris *may be avoided by showing that the creative process itself is not a free activity if by free we mean arbitrary, or unrelated to cosmic law. For the creative process in the human mind, the developmental process in organic nature and the basic laws of the inorganic realm may be but varied expressions of a universal formative process. Thus* World Perspectives *hopes to show that although the present apocalyptic period is one of exceptional tensions, there is also at work an exceptional movement toward a compensating unity which refuses to violate the ultimate moral power at work in the universe, that very power upon which all human effort must at last depend. In this way we may come to understand that there exists an inherent independence of spiritual and mental growth which, though conditioned by circumstances, is never determined by circumstances. In this way the great plethora of human knowledge may be correlated with an insight into the nature of human nature by being attuned to the wide and deep range of human thought and human experience.*

Incoherence is the result of the present disintegrative processes in education. Thus the need for World Perspectives *expresses itself in the recognition that natural and man-made ecological systems require as much study as isolated particles and elementary reactions. For there is a basic correlation of elements in nature as in man which cannot be separated, which compose each other and alter each other mutually. Thus we hope to widen appropriately our conceptual framework of reference. For our epistemological problem consists in our finding the proper balance between our lack of an all-embracing principle relevant to our way of evaluating life and in our power to express ourselves in a logically consistent manner.*

In spite of the infinite obligation of men and in spite of their finite power, in spite of the intransigence of nationalisms, and in spite of the homelessness of moral passions rendered ineffectual by the technological outlook, beneath the apparent turmoil and upheaval of the present, and out of the transformations of this dynamic period with the unfolding of a world-consciousness, the purpose of World Perspectives *is to help quicken the "unshaken heart of well-rounded truth" and interpret the significant elements of the World Age now taking shape out of the core of that undimmed continuity of the creative process which restores man to mankind while deepening and enhancing his communion with the universe.*

RUTH NANDA ANSHEN

WORLD PERSPECTIVES

Volumes already published

XXVII ALFRED NORTH WHITEHEAD: HIS REFLECTIONS ON MAN AND NATURE Ruth Nanda Anshen

XXVIII THE AGE OF NATIONALISM Hans Kohn

XXIX VOICES OF MAN Mario Pei

XXX NEW PATHS IN BIOLOGY Adolf Portmann

XXXI MYTH AND REALITY Mircea Eliade

XXXII HISTORY AS ART AND AS SCIENCE H. Stuart Hughes

XXXIII REALISM IN OUR TIME Georg Lukács

XXXIV THE MEANING OF THE TWENTIETH CENTURY Kenneth E. Boulding

XXXV ON ECONOMIC KNOWLEDGE Adolph Lowe

XXXVI CALIBAN REBORN Wilfrid Mellers

XXXVII THROUGH THE VANISHING POINT Marshall McLuhan and Harley Parker

XXXVIII THE REVOLUTION OF HOPE Erich Fromm

XXXIX EMERGENCY EXIT Ignazio Silone

XL MARXISM AND THE EXISTENTIALISTS Raymond Aron

XLI PHYSICAL CONTROL OF THE MIND José M. R. Delgado, M.D.

XLII PHYSICS AND BEYOND Werner Heisenberg